FLAWS
in
PHYSICS

from : before and including *GALILEO*

through *NEWTON*

for example of a contradiction :

2nd law   *"acceleration is **in <u>the direction</u>** <u>of the</u> <u>force</u>"*   [action / result **similar**]

3rd law   *"every action has ..<u>an</u> **opposite** <u>reaction</u>"*   [action / result **different**]

(see page 41)

to : including and after *EINSTEIN*

Publisher : p.l.fairfield@mail.com

April 2016

Printer : www.think-ink.co.uk

ISBN 978-0-9935834-0-7

CONTENTS

Appreciation

I recall (with pleasure and gratitude) the help and guidance given to me by professor Reg Lawrence during fifteen years at the Queen's University of Belfast, as well as by his numerous colleagues, who all introduced me to much information of great interest.

Thanks also to *Ansgar Allen* of the University of Sheffield, for pointing out a misnomer, and for posting a review of my work :
   "Accomplished satire ; absurdly true."

PRELIMINARY EXPLANATION

Dear Reader

If you like to exercise your mind about **prominent historical and current matters of public interest**, you could enjoy reading *Flaws in Physics* (well, that is if you accept the author's self-assessment).

It applies the scientific method recommended by *René Descartes*. That is to isolate the essential features of *any* topic - by questioning each of them rigorously, so as to identify those that are relevant and to discard those that are not; then to arrange valid ones in logical order to make a coherent whole.

That could be elementary, or complex (with various simple parts interconnected [everything I write here is simple]), instead of being complicated, which unfortunately happens in many cases.

Herein you can discover that **renowned scientists contradict others as well as even themselves** ; and how they can confuse you too. They state their views as if beyond argument. For example, nobody seems to have noticed that *Newton* contradicted *Galileo*.

Again, I do not debate *whether* man went to the moon - I question *how*. If you think my answer is wrong, please tell me *why*.

Of course the expositions here might appear impertinent when you recognise that *we* can enjoy the advantages that technology has brought about.

Yet *we* are not privileged by having our minds cluttered with information of doubtful value, as reviewed here. Clutter confuses the public - who are induced to believe, rather than to meditate (most people don't want to).

Thinking is a skill that has to be learned.

Then with some effort, *anything* can be stated simply : *Newton's* first two laws of motion (see page 40) are good examples.

Although Cartesian technique was well used by many distinguished persons before now, *Flaws in Physics* exposes several widely-accepted complicated conjectures, in different fields, that contain elements reminiscent more of mythology than of science.

Physics is the branch of science concerned with the properties of matter and the relationships between them.

Yet do keep in mind that my use of the word

'science' is in the sense of knowledge organised in a systematic manner (or should be), and remember that myths too are similarly formed into structured units. Remember as well, that the shortcomings I expose do not invalidate the numerous beneficial aspects of the various sciences.

In the following pages I examine the general knowledge of physic. It sometimes requires **unusual effort of thought** (*"cogito ergo sum"* = *"I think, therefore I am"* : *Descartes*).

Thus I contacted academe for comment. That was almost like trying to explain *Massenet*'s sublime '*Meditation*' to a deaf person.

Nobody could show any flaw.

Out of the several hundred academics I contacted -

one agreed that there was no flaw ;

from five there were doubting responses, to whom I suggested their mistakes ;

those who continued evaded admission, by suggesting either that I should read authors whom they named, or that any hypothesis is not valid unless supported by mathematical analysis (even though they had not done so themselves), and that they could not accept any proposition

that had not been 'peer-reviewed' (such 'peers' fill many journals with many vague articles that are presumably read by peers) ;
one replied - about Galileo - that I made an excellent point, and another - about Newton - that I was correct.

See also my article 3 (3) 2007 *'Galileo's Mistakes'* in the *Journal of American Science.*

Since I am acutely aware of not being exempt from making errors, I have perused this paper very carefully, in the hope of excluding any, and in the knowledge of the great difficulty for testing one's own self-assessment.

I do know that people challenged in any science, tend instantly to leap to self-defense of invented thoughts, without keeping to the point raised. I also have to admit that - where I say that nobody explains (in the relevant topics) - there might be explanations, but I have not come across any during several decades of widespread reading and listening.

Therefore, this is a fearless challenge to Physics and Astronomy Departments around the world. 9

However, *I promise to acknowledge faithfully any reasoned comment* to me, at :

davisonb0@gmail.com

Welcome to a Cartesian world where experience can be a tragedy to a person who feels ; while it might be a comedy to anyone who thinks.

Bruce Davison, B.Sc., D.G.A., Ph.D

PARTURITION

There is currently a choice among several universes, including those on offer from beyond black holes by modern physicists.

However, three of the universes (lettered A, B, and C below) might be one and the same - as follows.

A　　　　*A Big Bang*

From nothing

A suggestion has been advanced by scientists that there was a primordial explosion - about 13.7 billion years ago - from something infinitely small.

Isn't that 'nothing' ?

Any explosion of any matter in empty space would logically create a hollow extending sphere, because there is no force to stop it (apparently in those days there were no black holes with infinite gravity).

Thus such a sphere would have an increasingly-thin crust.

Despite the infinitely-small size, nobody

knows how the little bits of nothing from the outburst managed to form themselves into billions of huge celestial bodies - especially as there was no other force to do that.

B   *A Creator*

More Imagination

About six thousand years ago several imaginative theorists anonymously ordered "let there by Heaven and Earth" - which thus were as they are now.

Another order was "let there be light" - and there was light, as well as mornings and evenings, even though the Sun wasn't made until several days later.

A further order (among many other contradictions) was "let there be plants" - which thus were, even in the absence of solar warmth and rain.

C   *An Hypothesis*

Randomness

About 20 years ago, a proposition was advanced that the Universe might once have contained myriad dispersed morsels of matter moving randomly.

The morsels could have contained indestructible innate force, causing that movement and the formation of our Universe. You can still see meteors ('shooting stars') doing that.

That force is likely to pulse, and mutually to influence all morsels within range.

At the outset (and for remnant morsels today) the extent of pulsing was probably as little as that which just caused movement - in line with absolute zero Celsius, at which the morsels would be in their lowest possible state of activity.

CONSTITUENTS

A   ***Atoms***   - from time immemorial.

Smallness again

These are too small to be seen (nobody seems ever says how they can be identified).

They must have outsides (because - except for a tiny nucleus - they are empty inside).

Nobody ever says how to get through the outside, in order to know - only within the last century - that there is a nucleus inside.

Electrons   - from 1897

Infinitesimally-small ***electrons*** circulate around the nucleus (how they do that is not explained).

Nobody ever says how to identify an electron.

Nevertheless, it is alleged to have a radius <u>one</u> ***divided by*** 1 000, 002 817 940 285 *metre* (presumably when one was captured alive) !!

Despite that measurement, the nearest they have got to being identified is that they are alleged to cause the sparks that can be emitted from an electrode, in certain circumstances.

Electrons are alleged to be negatively charged (presumably with 'electricity') though nobody explains what 'negative' might mean, nor how to identify it.

Electricity is alleged to be 'carried' along conductors by electrons.

A definition

Note that the definition of *an ampere is the constant current* (presumably of electrons - and apparently by frequency, rather than by amount of electricity [which seems to be measured by volts]), *that, when maintained in two parallel conductors of infinite length and negligible cross section placed one metre apart in free space, produce a force of 2 x 10 to the power of minus seven newton between them.*

The definers do not say where their conductors may be observed.

Anything so small as to be 'not worth considering', can hardly be called scientific, because a conductor must have some material for the electrons to flow along.

Therefore *Newton*'s principle of reduction with distance from source would reduce the force to zero long before the alleged infinity (unless the scientists went along at the speed of light, to do the maintaining).   15

Protons   - from 1918

The nucleus of a hydrogen atom (the <u>first</u> element ever discovered, more than two centuries ago - long before modern atomic theory was imagined) is presumed to be a <u>single</u> **proton**.

Nobody can say how a proton can be identified (not least when protons are alleged to be in groups, now in sequential numbers up to 118 - with their sequence being uncannily in line with their 'discovery').

Protons are positively charged (presumably <u>also</u> with electricity, but protons apparently do not 'carry' any electricity anywhere).

Nobody has said what 'positive' might mean nor how to identify it.

Neutrons   - from 1920

Also part of the nucleus (except hydrogen, for some unknown reason) is a merely-postulated **neutron** - which is without charge (how to identify that and its host, is not explained).

However (contrary to all evidence) "*like* is often alleged to *repel like*" - so neutrons allegedly keep protons from repelling each other (sometimes rotating, sometimes not - both unexplained) when any electron slopes off to cause sparks, for example.

Now neutrons are alleged to have greater mass than protons - so the larger body can easily be made (somehow) to push a neutron out of a nucleus - without explaining where the lonely one goes to.

Moreover, radioactive decay (also known as nuclear decay OR disintegration, OR radioactivity) *is* the process by which a nucleus, of an *unstable* (what's that ?) atom, *loses* energy by emitting ionising radiation that carries (from where ?) enough *energy* to free electrons from atoms, OR from molecules.

Quarks   - from 1932

However, allegedly there are three infinitesimally-small *quarks* inside any proton or neutron.

How the outsides of the two nuclei can be identified is not explained, nor how to achieve penetration, nor how to identify the tiny tots.

Despite their littleness, they allegedly exercise the greatest force known to scientists - which is *given* as the reason for the immensity of atomic explosions (explosive charges surround a material whose pulsing agitation has been increased in advance).

That complicated, alleged, fission is undertaken despite the ease of pushing (mentioned in the penultimate paragraph of the previous section about neutrons).

B   ***Creation***   - from circa 6,000 years ago

The Heaven and Earth might have come about from *"waters divided from waters"*.

C   ***Morsels***   - from circa 2,000

Matter

A visionary bishop, called *Berleley*, in 1709 suggested that matter and energy are one and the same thing. That could be true because our awareness mechanism only registers influences (thus where, but not what, they come from).

Energy is defined as "the capacity of a body or system to do work" - thus necessarily involving force, of which we can be aware (see also next section).

An imaginative theorist, called *Einstein*,

in1915 was acclaimed by peers to have originated *Berkeley's* very thought. It now took the form which proclaimed that matter can be changed into energy (why not into force?) and vice-versa.

From that came the formula $E = m.c_{squared}$.

That does not conform to the rules of simple arithmetic :

multiplication is a convenient statement of the number of times something is added ;

thus anything may only be multiplied by a number ;

the reason for using any multiplier must be explained, but no reason has been given for the suggestion to use $c_{squared}$ (whatever the figure might be) ;

the result of multiplying can only be the same as the multiplicand - so mass multiplied by any figure would be mass, not something called energy.

Force

Logic suggests that this is likely to be pulsing in all directions.

It is also likely to be mutually influencing all other morsels within range.

19

Simultaneously, the pulse is likely to be strongest in the direction of a body's movement.

Scientists at CERN say that their magnets can make particles move in a circle without the magnets themselves rotating, whereas force can act only in a straight line from a magnet to any particle it attracts.

Extent

By observation, too, the greater the number of pulsing influences, the greater is the extent of force exercised.

However, the extent of one body's influence would reduce with increasing distance from source - as everyday observations demonstrate, such as the centre of a steak's being rare when the outside is already grilled.

That further illustrates that morsels take time to react (which I call ***reaction-time***) as the morsel nearest a source of influence has to adjust to it, before that morsel can influence contiguous morsels, and so on sequentially.

Moreover the greater the extent of force, the less time is needed to achieve a similar result.

Space

This can be observed because of a multiplicity of positions for matter (or for sources of influence, if you prefer).

Time

The <u>series of movements</u> in space, all involve intervals that we can measure relatively, and can <u>call time</u>.

Measurement

Since everything is relative to something else, the best measurement of anything is that which relates to the least change in the elements used as criteria.

GRAVITY

Tendency to move

The random movements of morsels would often result in the coming together of two morsels.

Then their mutually-pulsing force would keep them together and cause them to move in one direction, instead of the former two.

The arrival of a third morsel would do the same, and - depending on its angle of arriving - could additionally cause rotation (**rotation** may be defined as turning 'on an <u>internal</u> axis', and **revolution** as 'on an axis <u>external</u> to the body').

Further supplements, with continued mutuality, would cause a spherical formation.

The two original morsels would be at the centre of the sphere, and be subject to the cumulative force of all other morsels - each acting mutually on neighbours, and thus towards the centre.

Such centripetal force is called ***gravity***.

Effect

The greater the force acting on a morsel, the greater the agitation of its pulses.

This can be recognised, by our awareness mechanism, as ranging through what we call 'heat' and 'light'.

That is why the Sun is warm and bright.

The 'nuclear reactions' of physicists is imaginary nonsense - if fission did occur [without their explaining why] how could consequently-necessary fusion follow, continuously for billions of years ?).

The gravity of any group of morsels would - by logical extension - mutually influence every other body within range, however distant.

Attraction

That concept of distance is incorporated in *Newton's* **law of universal gravitation :**

"*every point mass in the universe attracts every other point mass with a force that is directly proportional to the product of their masses and inversely proportional to the square of the distance between them*" ($g = m.m / d_{squared}$) :

That does not conform to the rules of simple arithmetic -

anything may only be multiplied by a number ;

the reason for using any multiplier must be explained, but no reason has been given for the suggestion to use mass (in 'm.m') ;

23

the result of multiplying can only be the same as the multiplicand - so mass multiplied by any figure would be mass, not something called gravitation ('g') ;

anything may only be divided by a number ;

division is a convenience for stating the number of times something is subtracted ;

thus the reason for using any divisor must be explained, but no reason has been given for the suggestion to use distance squared ('d$_{squared}$') ;

the result of dividing can only be the same as the dividend - so mass divided by any figure would be mass, not something called gravitation.

Calculating causes

Since influence seemingly acts globally, the magnitude of the influence of **one body**'s gravity at any point would probably be inversely related to the respective areas - on the one hand, of the sphere encompassing that point, and - on the other hand, of the sphere from where the gravity emanates.

Note that the area of a sphere is related to the square of the radius.

The net gravity of **two bodies** would be the *difference* between their gravities thus

24

calculated (because they act in <u>opposite</u> directions, hence the subtraction, <u>not</u> *Newton's* multiplication).

Accordingly, either body could only move towards the other if influenced by a force <u>greater</u> than its own - whereas observation and logic suggest that the difference is zero at some point *between* any two cosmic bodies, such that neither can move towards the other (that is why *Le Verrier's* alleged disturbance in planetary orbits is doubtful).

More doubt

Thus it is not surprising that *Newton* wrote (about gravity) : *"whose cause is what I do not pretend to know"*.

Indeed, Galileo had already admitted that the *cause of falling "is not a necessary part of the investigation"*.

Apparently, up to date, no scientist (unless you admit the present author as one) has ventured to supply the lack by Galileo and Newton, though the dictionary defines gravity as *"the force of attraction that moves or tends to move bodies towards the centre of a celestial body"*. Note that the whys and wherefores of the force are not indicated in the definition.

Weak carrier

Certainly there <u>is</u> universal gravitation, but there is <u>not</u> universal attraction - because a body's reduction of gravity with distance usually results in an influence that is not greater than the gravity of most other celestial bodies.

Nevertheless, scientists say that there is a particle, called 'graviton', which is <u>known</u> **(despite not having been found** !) not to have mass, yet carries gravity anyway !!

They even claim to measure the strength of gravity in an atom, but don't explain how, nor in what way, they distinguish it from three other forces they have imagined.

BLACK HOLES

These dark cavities exist by reason of immense gravity (it's impolite to ask where such gravity, and the 'anything' that allegedly falls into the holes, come from).   Even photons (carrying light, which allegedly is 'white' when it is 'visible') couldn't get out from a black hole - which is purportedly why such holes are black.

Contradictorily, 'radiation' could escape, and *"escape velocity exceeds the speed of light"* even though *"nothing can travel faster than light"* (many scientists say).

Indeed, the *gravity* of a black hole becomes so incredibly strong that the vacancy gets smaller and smaller until it becomes *"infinitely tiny"* (somewhere in the hollow sphere that a big bang must create).   Isn't that '*nothing*', or should one say *"totally collapsed matter"*?, as scientists allege, whatever that might be.

It's also impolite to ask the 'why and how' of that smallness, as well as what attributes it has that enable it (and its alleged rotation!) to be measured at any stage.

Don't black holes thus remind you of the big bang in reverse, and accordingly of physicists standing on their heads ?

Do those then *"collapse under their own weight"* ? (as the equally-weird holes are alleged to get up to).

Singularly strange

Do you remember that gravity is the influence of grouped matter's mutual force ? If yes, you might well think that - since black holes don't have any matter - they therefore can't have any gravity ! Neither can they have any photons to be kept in (nor to hover) - because, in the absence of matter, there can't be any *material* electrons to vomit them.

That's because at the centre of a black hole there is allegedly a *"singularity"* - where the density and space-time are infinite !!! Well, there is if you believe *Einstein's* general relativity, poor soul.

Isn't science-fiction astounding ?

That might be why photons could impossibly hover (with finite *speed* ?) at the *"event horizon"* of a black hole - allegedly omnipotent thus attracting ALL matter !

GALILEO

Falling

The investigation that Galileo wrote about was an experiment to test whether large bodies fell (under the influence of gravity) more quickly than small ones - as *Aristotle* had argued two millenniums earlier (without explaining), followed by other 'scientists' ever since.

When Galileo did let two objects fall from the top of a tower in Pisa (which was conveniently leaning) he observed that they arrived at the ground apparently simultaneously.

Then - in order to be more scientific - he rolled different-weight same-size balls down a long graded slope. He measured the time too :

first by collecting flowing-water, relating to several equal distances, and

secondly by counting the number of grains of sand that balanced the weight of water (the random distribution of different sizes in grains, permits taking them all as the same). Samples were taken at successive units of distance.

The number of grains at each *period* was successively equal when the number of cumulative corresponding *distances* in units was measured *at* 1, 4, 9 etc.

Interesting increments

Now logic suggests that a unit of distance is induced by a unit of force. That unit is the force **in _each_ individual case** needed to change the relevant body from rest into movement.

Note that, in line with reaction-time and *Newton's* second law, a newton would take *longer* to start movement in a mass of two kilograms, *than* in a mass of only one kilogram.

Accordingly, motion is not caused by the one unit of force applied during start-up reaction-time. Therefore the first unit of distance is related to two units of force.

Inertia

There is also reason to think that, after a unit of force causes movement (if there is no external other force), it will continue (as *inertia*) to induce a unit of distance, during successive intervals.

Thus *Galileo's* distance-series starts at 1.

Then the distance 4 comprises that 1 + *1* (from applied inertia) + 2 from new application of

two (such as in kilograms, which people tend to compare alone), thereby making two immense numerators infinitesimally different from immense denominators which are the very-same *mutual* gravity.

Therefore the difference in fractions (thus in the time for the reaction) would be '*very* difficult' to notice. You might as well try to see how much the ocean rises when you go paddling. I know, it rises over two feet !

Effect of distance

Another difference that would also have gone unnoticed because of its minuteness too, is the concept of increasing influence on approach to source : the extent of gravity at the top of Galileo's experimental slope would be less than at the bottom.

However, the measure of these two differences is so small that they can reasonably be ignored for all practical terrestrial purposes, though they might need to be accounted for on a cosmic scale.

GEOCENTRICISM

Orbit

After the Heaven and the Earth were got going by the imaginative theorists, other theorists alleged their observations to confirm that the sky orbited the world.

That's geocentricism.

At that time, anyone who was anybody lived, for example, where a view to their left could have been a high eastward hill (as *Shakespeare* noticed - millenniums later - the Sun had the habit of creeping over it every day).

Then it went over, say, a mountain in the observer's view to the front (which would be the south) before continuing and setting behind, say, a forest to the right (which would be the west).

Relativity

Observably, the hill, the mountain, and the forest, did not move (either closer to, or farther from, each other) and the observer's vantage point was equally stationary relative to those three fixed points - whereas the Sun could be seen consecutively above them daily from east to west.

The stars too could be seen at different times above the fixed places observed.

Such evidence suggested that the Earth was not moving (though Aristarcus did propose a contrary case that it did - to which nobody would listen [I could sympathise with him]).

Proving the rule
An exception to the solar movement occurred when an experimental physicist stopped the Sun, such that it *"hastened not down"* for almost a day.

Some non-experimental theorists also supported the earlier imaginative ones by alleging that the Earth had been created *"fixed and immovable"*.

Indeed, a renowned philosopher, called *Aristotle* (who led a large clique of physicists, including *Ptolemy*) firmly alleged such fixity to be a fact.

Further proof was drummed up by burning alive anyone who opposed geocentricism : in the year 1600, *Bruno* for example, was given such heat-treatment, but *Galileo* retained his cool (after a trial in 1632) by muttering nonsense daily for many years until he died - even

35

though he had disproved geocentricism by observing moons orbiting Jupiter.

HELIOCENTRICISM

Rotation and revolution

Yet the appearance (of the Sun's crossing the sky from east to west, in a period called a day) would be the same if the Sun was stationary and the Earth turned from west to east.

That's the same for the stars [with *unchanging* so-called 'trajectories' (except 'stationary' *Polaris*)].

Their daily 'advance' on the Sun (additionally) would be accounted for if the Earth was orbiting the Sun during the period that is called a year.

The disappearance of many of the stars for half a year would be accounted for by the fact that stars can't be seen in daylight - which for the hidden stars would be from one end of a solar-terrestrial diameter, and they could be visible from the other end facing the dark during the following half-year.

Inclination

Furthermore, the biennial perambulations of the Sun to and fro (thus *changing* 'trajectories') along part of the terrestrial horizon - as the

Earth rotated daily and revolved yearly (note that all the farthest stars of astronomers have the Sun as the central point - which is heliocentricism) - accords with a terrestrial axis *leaning* almost permanently (little force to change) towards one particular star (which is *Polaris*, at present).

Size

The enormous extent of that system can be divined by observing :

<u>first</u>, that the great curvature of Earth can be deduced by distant snowlines lower than proximate ones, as well as by the gradual disappearance of moving ships below the horizon ;

<u>secondly</u>, that the large size *and* distance of the Moon is evident from its ability to light vast areas of Earth, and to move a colossal volume of ocean-tides (plus atmosphere and *'leaning'* continents) ;

<u>thirdly</u>, that the visualised central radius of a 'half' Moon lit by the Sun seems parallel to one's view to our star - which must therefore be at immense distance and thus be incredibly big ; and

fourthly, that the apparently tiny stars are bright and thus probably suns too, so must therefore be enormous as well, and phenomenally distant - thereby suggesting a vast universe.

NEWTON

Impeccability

The **first law of motion** was restated by *Newton* : *"if an object experiences no net force, then its velocity is constant : the object is either at rest (if its velocity is zero) or it moves in a straight line with constant speed (if its velocity is nonzero)"*

Having got that law from China, he probably got the idea of straightness also from *Descartes* and others.

That is sometimes called the law of **inertia** because of the obvious lack of changing state (*Newton* reputedly held that continued movement does not need a body to have force in order to do so - which raises the question, unanswered, as to where force comes from when it is exercised).

Note, that rest is when two bodies do not move *relative* to each other, even though they could be moving together.

Contradiction

The indisputable **second law of motion**

includes : *"the <u>acceleration</u> of a body is ... in the **direction** of the net force ..."*
    force --- > --- > **result** (<u>same direction</u>)

The **third law of motion** states *"to every action there is always opposed an **equal** reaction directed to **contrary** parts"*
    force --- > < --- **result** (<u>opposite direction</u>)
[could also be : < --- --- >]

<u>Yet</u> if that 3rd law was true :
a free kilogram could **not** move at all, even a metre, when affected by a force of one *newton* (see its definition) - <u>it does move</u> ; and
  compare two equal locomotives trying to pull or push each other to opposite terminuses - <u>they do not move</u>.
Thus the **third law is invalid**.

False proof
  Some physicists suggest that (for example) a space rocket illustrates that the third law IS valid - on the grounds that the exhaust emission can be observed to be in the opposite direction to the rocket's flight.
  They simply overlook the fact that the

41

explosion of a fuel is multi-directional, and

first therefore would not induce any movement if acting within a firm enclosed chamber ;

secondly a chamber open downwards would enable an explosion to occur - with exhaust gas emitted impotently in that direction ;

thirdly then the horizontal effect on the firm sides of the chamber would also be impotent ; and

fourthly the vertical upward effect of the explosion would push the rocket <u>in the same direction</u> as that effect.

Ellipse

Newton also stated that the laws of motion confirmed *Kepler's* <u>ellipse</u> in that astronomer's **1st law of planetary motion**.

Yet solar influence (aiding a planet's orbit) ***increases*** approaching perihelion (near), and ***reduces*** approaching aphelion (far) - so those *different* forces could hardly cause the *similar* opposite curves that an ellipse has.

PHOTON

Remarkable runner

Encouraged by *Planck* - who reasonably hypothesised that force ['energy' actually] is composed of packets, called quanta) - *Einstein* exclaimed : *"imagination is more important than knowledge"*.

Consequently he imagined that light comes in such packets too (which came to be called 'photons').

As a result, the imaginary photon ran riot :

sometimes it **was** light, sometimes it **carried** light ;

otherwise it carried, or was : microwaves, heat, infrared or ultraviolet radiation, radio, x-rays, or gamma rays ;

it can move as if it is matter (which can have '**speed**') ;

photon's alleged speed in space is finite and is the greatest possible - but *somehow* it can be *stopped* ! (and thus become 'virtual') ;

in addition, photon **is** a ray ;

otherwise it carries radiant energy.

Mass is a must

Mass **is** a 'quantity of matter' and therefore has weight ;

Einstein argued that he could evaluate the weight of a photon by letting it out of a box ! ;

nevertheless some physicists say that photon has <u>no</u> mass ;

that conveniently overlooks *Einstein's* formula $E = m.c_{squared}$ - if m is zero then E would be zero - whereas they <u>also</u> say that photon IS energy (even *"pure energy"*), sent to our eyes by objects.

Nothing's the matter

They also allege that energy can exist without matter ;

note that - according to *Hawking* and many other grazing mythologists - energy *"fills space"* by *"escaping"*, <u>as well as</u> totalling to *"nil"* ;

photon **is** also a particle - which is defined as *"matter"* - pardon : *"a physical entity"*.

Wavy wobbling

Otherwise it **is** also a wave (but not when 'virtual', one could suppose) ;

44

as such it can alter the trajectory of an electron inside an atom, (to create the frequency spectrum) ;

in electron-microscope-lenses, wavy photon can encourage electrons to create 'magnetic fields' ;

waves involve distance, so photon has wavelength ;

in fact it has frequency (which is a number of pulsations in a certain time, regardless of distance) ;

as well as being a wave (waves are *curved*), it travels in *straight* lines (how could a piece of matter do otherwise than travel in a line? like the spokes of your bicycle wheels) - therefore it must be non-existent in some places throughout space (like the empty expanses between the spokes) ;

nevertheless, it *is* <u>everywhere</u> in space from *any* <u>one</u> source !

SPACE-TIME

Bendy boffins

Einstein's General Relativity (page55) is made to allege that space and time form one entity ;

additionally, such *space-time* is curved ;

however, photon's straight lines aren't when the gravity of stars bends them ;

yet the Sun has a halo, which refracts light - which is hardly 'bending' *by gravity* ;

furthermore, the 'photons' of an alleged radar signal would not only be bent going, but also would not (at 'c') be able to rebound in the short time of observation - from the astronomically distant star, as astrophysicists sometimes claim it does ;

what's more, space-time is **not measurable** ;

time **can** be measured (only separately, by noting displacements, or recording pulses) ;

space can be measured too (with objects relative to each other), in conformity with *Euclid's* geometry - which **is** consummate (despite allegation to the contrary by voluble physicists) ;

they 'illustrate' *Einstein's* case with a diagram

of a net showing a body causing <u>curves</u> down towards where it depresses the net - just *you* try that, and you will see only <u>straight</u> strings where *they* show curves ;

you'll also be able to observe that the body *<u>pushes</u>* the net, whereas gravity *<u>pulls</u>*.

Speedy sports

Nevertheless, donnish astronomers *say* they can tell which photons come from the farthest stars (seen *<u>where they were</u>* as those 'carriers' allegedly *<u>started out</u>*) ;

thus the stars must have moved faster than light, to have arrived so far (13 <u>units</u>) so soon (0.7 <u>units</u>) after the big bang ;

in any case, how could photons coming from the farthest stars be identified, if their trajectory is *not* straight ?

HEAT and LIGHT

Agitating actions

Influence in morsels, accumulates in each towards the centre of any group. The greater the accumulation the greater is the agitation in each morsel - which can be called pressure.

Greater pressure may be recognised - by our awareness-mechanisms of touch - as **heat**. Then, as pulsing frequency increases (that's more agitation in the same time) its influence can additionally become **light** to our awareness-mechanism of sight.

Thus different mechanisms react differently to the same influence.

That too is relativity.

Environment

Note that heat can not *"escape into space"* as scientists say it does, because it is not a substance. It is simply a rate of pulsations as influenced by the environment.

Such pulsations are why the immense Sun shines warmly (see 'effect' page 22).

Naturally the more it thus affects part of Earth,

the more that part is heated. That is what happens when the slow-moving Moon is 'behind' Earth and above a 'higher' latitude than the Equator, with the result that the 'opposite' part of the southern Pacific is less pulled to the west by lunar gravity, and is therefore more heated (also with less replacement cold water pulled from the Antarctic Ocean) ;

that greater sea warmth is called *El Nino*.

Cause and effect

If you use a hot-air dryer on your hands after washing them, that air causes the dampness to evaporate - by increasing its pulsations in one direction to cause its movement : that results in reduced pulsations in other directions and the influence is transmitted to your skin, with the reduced pulsations being registered as cold.

In addition, force is the cause of motion, related to the rate and extent of pulsation, so increased agitation can result in volcanoes on Earth, and in solar flares spurting - note that they fall back, whereas physicists allow protons to keep moving out.

GLOBAL WARMING

Delayed diddling

Scientists say that, soon after the big bang, the Universe cooled.

That is impossible because heat is only our awareness of agitated pulsing in <u>matter</u>, and the big-bang-particles were without any external force to change their state ;

the increasing space occupied by them can not be part of the measurement, because emptiness has no temperature.

Then 'cooling' ceded to the dire threat of 'warming' on Earth, in recent years ;

that is now ceding to 'climate change' ;

the Pharaohs in Egypt had that change too from time to time during their three millenniums of rule - but apparently <u>not caused by pollution</u> (as global warming is alleged to be nowadays) ;

yet modern *dire* threats about warming are ceding to *beneficial* effects on agriculture ;

carbon dioxide [CO_2] is ceding to methane ;

both of these are letting moisture take over from the 'glasshouse effect' (even though it is <u>cooler</u> under clouds), without reference to anything like fixed glass and walls (which frustrate air-movement, and so contribute to the agitation we call heat).

Alleged proof

An attempt to justify the claim, that warming was happening, was made by a professor from Switzerland several decades ago.

He reported in an academic paper that he had studied ice cores and tree rings [both in his own country] to reconstruct the climatic history of Europe over the past one thousand seasons. He did not disclose how he determined temperatures from them, nor how they served for the whole of that area.

Yet as he developed his theme, that did not prevent his blithe extension from Europe to the northern hemisphere, then to the whole world - without any supporting data for the extension.

Meanwhile there was no known change in the Sun's ability to heat.

Moreover, he did not admit that CO_2 is heavier than air, even though the core would have shown that it falls to earth - whereas other physicists were claiming that it formed a layer higher in the atmosphere.

Temperature

Those scientists also claimed that they recorded a one degree Celsius rise since before the industrial era (when there would have been fewer CO_2 emissions than today).

They now have about eleven thousand weather stations around the world, from all of which to record.   Yet there were certainly very few stations before 1760 with which to compare, and they did not say whether they compared only with them, as a scientific study would require.

Neither did they say how many of their readings were made in cities - which *are* warmer nowadays (the first coke-fired furnace was instituted in 1710), because we heat our towns but not the vast areas of surrounding countryside.

Indeed, the calibration of early thermometers (mercury was first used in 1714) was not verified, because it was not known then that glass stabilises during about a year after manufacture - so the few early readings could not be entirely reliable.

More unreliability

The weather savants also claim that the sea

level is rising because polar icecaps are melting, even though the scientists have no fixed base for any such measurement (the Earth's surface is always moving because of lunar gravity) ;

icebergs are often portrayed as an example of that melting, even though they have always existed, having been induced by gravity (glaciers and packed ice move by virtue of the weight of snow and ice on them, which can therefore break off into the sea as icebergs - have you heard of the *Titanic* arguing with an iceberg in 1911, and the *Hannah* with disastrous results likewise in 1849 ?) ;

indeed, polar icecaps are imported on wind *below zero* temperature (the Sun's influence is at or near right angles) ;

there is possible terrestrial axial wobble (by interaction of lunar and solar planes, possibly accentuated by localized variations in quantity of ice resulting from prevailing winds) as evidenced by seafaring charts ;

high summer pastures have not noticed significant variation during more than a century;

on the other hand social problems have <u>recently</u> been caused by the coldest dormant season in living memory, the greatest cold on record has been recorded in Siberia and the Antarctic, and the coldest in Paris since a century earlier, with the freezing of the sea and rivers from time to time and from place to place ;

there are also free holidays called conferences.

EINSTEIN

Relativity

Einstein imagined weirdly that *"the interval of time between two events or their simultaneity is not absolute"* ; and that *"observations have different values for different observers"* (he said!).

In support of that fantasy, *Einstein* is attributed with a 'thought experiment' in which an observer in the middle of a moving train passes another observer on a station platform at exactly when two flashes of lightning simultaneously (how determine the sameness ?) hit the front and back of the train respectively.

His argument (purportedly) is :

<u>*first*</u>, since light has *"finite speed"* - (note that, apart from reaction-time in intervening recipients, the movement of any matter does not affect any influence that it issues) - the platform observer <u>*will see the flashes as simultaneous*</u> ;

<u>*secondly*</u>, the moving-train observer is approaching the front 'flash', and is withdrawing from the rear one, so will 'see the flashes'

respectively before and after the platform-observer does - and will therefore determine that *the flashes were **not** simultaneous*.

Ignoring facts

Scientists do not want to be told that *Einstein* did not take account of all relevant facts.

Two of them (whose difference the physicists ignore) were :

1 the platform observer was influenced at *one* point in time and at *one* point in space relative to the *other* time and places of the flashes

2 the moving observer was influenced at *two further* points in time and at *two further* points in space.

Note that the <u>one</u> word "flash" is therefore not representative.

Note also that such observations could be *independent of observer-movement* (the movement of the train was irrelevant, because three observers could have been placed at the points where the influences were received).

Flaws

Thus *Einstein* did not distinguish :
<u>on the one hand</u>, the ***issues*** made by the

flashes, and <u>on the other hand</u>, the **receipts** of their various (different) *influences* (at three different places, and times as well).

That is because observers at different distances from a source (of light or gravity or otherwise) could <u>not</u> experience similar influences, when these take different lengths of time to travel different distances through atmosphere which is subject to reaction-time.

Consequently, *Einstein's* conclusion was sheer nonsense.

This analysis strongly suggests that 'simultaneity' IS absolute.

Note too that *Galileo* had already misled people by **making out** that a sailor would 'see' a stone falling <u>straight</u> down in front of a mast, while an observer on shore would *'see'* a <u>curve</u> (caused by momentum and by *visualising* the stone's position when starting and again on arriving).

Note that both the sailor and shore-observer could have visualised both phenomena, but *Galileo* did not refer to that possibility.

Tardy time

Unabashed, *Einstein* reputedly argued that - as speed approaches 'c' - *time* passes more slowly (without saying how time could be measured, to support that nonsense).

He also argued, foolishly, that <u>*time*</u> could alter, when two <u>allegedly</u>-identical clocks were found to read one phenomenon differently. Now force and matter might or might not be one and the same thing, but an *instrument* (such as a clock) and *the phenomenon it measures* (time) are *always separate*. The method of measurement can be flawed, but physicists expect their audience to accept that the two clocks were not affected by extraneous influences.

That might be why the *magi* (travelling by stealth at night !) arrived simultaneously with the star that "*went before them, till it came and stood*" over the West Bank.

According to *Matthew* the magi were wise men.

SPACE TRAVEL

Up and down

Now sufficient force **can** move matter in the direction (upwards) opposite to gravity's attraction (downwards).

In logic, an upward movement could continue only so long as the matter exercises a force *greater* than the mutual influence of gravity at the point of action.

If there is no such force, the upward-moving matter will eventually be influenced to change to downward. Everyone can have observed that by throwing up a stone - gravity acts on it like the brakes on your bicycle, which cause it to stop unless some greater force (like pedalling) continues simultaneously.

Neither up nor down

If an <u>external force is applied</u> to the smaller body - such as a propelling fuel upwards - it could continue until the propulsion stopped where the body force and that of gravity equate.

Then movement would stop.

Of course if the gravitational force is rotating

(such as the Earth's) the net force could make the small body either geostationary, or enable it to orbit terrestrially (like the residual now - of an original momentum such as the Moon's at right angles to Earth's gravity).

Thus the Moon, and the planets, probably came from outside the solar system, at various times. That is further suggested by craters on the Moon's surface 'facing' the Earth (which, as the considerably bigger in a couple, would otherwise have attracted the supposedly-causing meteoroids).

Note that terrestrial meteors do not fall perpendicularly.

Propellant

Bearing in mind the braking effect of gravity from any body, and the probability that *immense* Jupiter is influenced by solar gravity, a *little* spaceship couldn't get that far without using a constant propellant (to overcome solar attraction) - so it's even less likely, ever, to *"leave the solar system"* (as astronomers blithely claim).

Yet many writers say that space-craft can be given a speed - ***"escape velocity"*** - leading to *"**cruise mode**"* needing no further propellant to continue thus anywhere in space, apparently unaffected by immense solar centripetal gravity.

Remember that even speedy comets are brought back from vast distances to keep orbiting the Sun.

Thus moonstruck scientists should realise that a *fuel supply would be necessary*, first to overcome Earth's gravity over a very long range; then to counteract lunar attraction for a considerable distance ; and subsequently to permit lunar orbit.

Lunar non-rotation

That the Moon does **not** *rotate* (*Galileo* reputedly thought it did) can be deduced - discounting libration (from solar gravity), which also indicates non-rotation - by observing that *a straight line never changes from the Moon's centre through a point on the lunar surface to the nearest point on the Earth's surface.* Yet that can be observed by anyone (well, except astronomers and *NASA* ! [do you rotate as well as revolve when driving on a traffic roundabout?]).

You could test for non-rotation :

first, by fixing a small sphere (representing the Moon) on the rim of a wheel : as the rim revolves around the hub (axis external to the sphere, equivalent to Earth) *the same half-sphere is always presented to that centre*, and all the surface of the ball is presented to an outside body (representing the Sun) : *that **is** what we can observe with our Moon*.

secondly, by installing a similar ball free to turn (on its own axis) and then rotating it once during a revolution : *the whole surface is 'seen' from the hub* 'Earth' while only one face is presented to the 'Sun' : *that is **not** what we observe*.

Regular recurrence

In addition, counteracting fuel would be needed for descent to the lunar surface.

Moreover, subsequent reverse of these would be required for getting back home - all needing fuel (some informed physicists say that *NASA* has not supplied relevant figures - thereby possibly avoiding moonshine, but maybe not avoiding visual effects for TV, nor possible hypnotism).

Also, ***planetary travel*** would need fuel to counteract gravity -

when **falling (*down*** towards the Sun) for 42 million kilometres to Venus (just imagine otherwise the vast acceleration if *Galileo's* formula applied) **;** or

when continuously **rising (*up*** from the Sun) for 778 million kilometres to Jupiter (needing two hours to exchange radio *sensitive*-control signals **!**), and beyond.

Aren't the astronomers' *"cruise mode"* (needs no fuel) **and** their *"gravity assist"*, just as much fiction as the *"anti-gravity coating"* (a substance he called *"cavorite"*) applied to the imaginary space craft of *H.G. Wells* ?

As for 'gravity assist' would pulling _towards_ Venus (but somehow not to the Sun **!**), **not** be followed by pulling _back_ ?

MORE IMAGINATION

The plutoed planet

Similar 'gravity assist' (by Jupiter), as well as that 'mode', were allegedly obtained for a jaunt to Pluto.

Now to control the craft near that vastly-distant small planet, as well as the orientation of equipment, would require more than eight hours (well, that's what the physicists require as electromagnetic transmission of photons) for radio contact - from there and back again, plus time to consider relevant information.

Nevertheless, both radio sets on the Pluto-jaunt were allegedly capable of emitting (and directing towards what would be equivalent to a pin-prick on a vast arena) such powerful signals that their reduction with the enormous distance did not diminish their effectiveness (in camera-manipulation, plus taking, sending, and receiving photographs as good as *artists' impressions*) even when 'behind' Earth for possibly twelve hours daily.

Fortunately - for avoiding the need of fuel - some 'cavorite' might be borrowed from Wells's

Garage.   That might overcome immense solar gravity during a cruise, of possibly more than 4,437,000,000,000 metres (its perihelion) *up* to Pluto, taking about nine years **;**

when *Buzz Aldrin* discarded a rocket (which apparently hadn't a cavorite covering), it went *"on its way in the direction of the Sun"* (that's *down*, as *you* probably would expect).

Curtailed cruising

An inquisitive person would be bound to wonder how spaceships could cruise without being restrained by solar gravity, when relatively-huge comets are not (because they are ceaselessly influenced by that attraction).

According to available evidence, a comet can round the Sun at many hundreds of times faster than the so-called 'escape velocity' of a space craft to leave Earth.

The probable reason for that superior speed is that solar centripetal gravity causes comet-acceleration (on its long approach to the Sun, after it first comes from outside the solar system), while solar rotation makes for a curved trajectory.   That combination leads to shooting away from the Sun as the increased momentum

65

exceeds the net gravitational force.

Subsequently, that which had been an accelerating approach, becomes a decelerating retreat as the two-fold solar gravity continues to act, but with decreasing effect as distance increases. Thus the shape of the trajectory becomes 'egg-shaped' (as with planets for the same reason) as the comet is made to return, from far beyond *Pluto's* distance.

Imaginary inventions

Those fuel-fantasies might be why *Einstein* said "*let there be a cosmological constant*".

There wasn't.

Like all so-called constants, it was just an invented solution where there was no problem (he admitted later - possibly having been misled by *Newton's* false "equal and opposite", which apparently makes all scientists think bizarrely that gravity needs a balance. They're not as willing, as *Einstein* was to admit contrivance).

Bloated emptiness

Now none of the yarns about space travel take account of the substances which scientists say fills the cosmos (called 'aether' by the Ancients).

Even *Rabelais* wrote *"nature vacuum abhorret"* (Latin had become the rage then, instead of Greek) ;

Descartes said *"reuguare ut detur vacuum sive in quo nulla plare sit res"* ; and

Newton (with economy of thought, if not of words in the latest rage, English) could not escape prejudice : *"That a body could act on another at a distance, through a vacuum, without the mediation of anything else, by, and because of, which their action could be transported from one to the other, is to me an absurdity so great that, I believe, no man who has, in philosophical matters, a competent faculty of thought, could ever fall into it."*

Later, physicists additionally filled space **simultaneously** with *"protons travelling at very great speed"* (even though, for example, solar flares <u>fall back</u> !) as well as with *"cosmic rays"*, and with help, of course, from *Einstein's "photons"* (as required by *Newton*), plus *"stellar dust"* and **"ice"** (thrown off by **hot** stars !).

The existence of dust is known (scientists say) because Earth '<u>runs into</u>' comet-trails ;

yet being tiny, would they not travel <u>faster</u> than Earth (which can't 'catch up' on them), or - having allegedly been denuded by 'solar wind' (imaginary ? in the 'vacuum of space' !) - could *not* even linger (because the 'wind' would carry them on) unless there was some other force (apparently there isn't) to stop the dust from obeying *Newton's* first law of motion (which obliges that continuation, or orbit, or falling) ?

At the same time the physicists <u>contradict</u> themselves (as usual), by adding that space is "*the greatest vacuum*".

As for comets leaving trails, note that they are sometimes described as immense balls of gas around a core, and that the 'tails' are illuminated in line away from the Sun - thus likely to be frozen matter, rather than the result of imaginary wind (in a vacuum !).

That might be why, when Einstein realised that a vacuum container could keep things hot or cold - he went off (to imagine) with his Thermos flask containing a cup of hot cauldron-juice and a ball of ice cream.

You probably realise that travel beyond the Earth's gravity is a lovely dream, and apparently useless.

MOON LANDING

Setting off

In a book that *Buzz Aldrin* wrote about the
Moon, he noted that *"the giant rocket"* -
"streaked (three hours after boarding) *from a
straight vertical shot to a gradually changing
angle of inclination"* ;

without explaining how   the angle changed
(an additional force to cause change is required
by *Newton's* logical first law of motion).

"After (another) *three hours, a third-stage
rocket engine* (as per *Newton's* first law) *took
the craft out of Earth's orbit -*   by *increasing
speed to nearly 25,000 miles per hour"* (40,000
kilometres per hour [kph] *"heading towards the
Moon"* ;

without explaining what force - (apart from
"throwing a switch") - subsequently sent that
rocket *"on its way in the direction of the Sun"* ;
nor why.

Travel time

Therefore   -   since the Moon's greatest
distance is given as about 400,000 kilometres -

the reader could expect the journey to take up to ten hours. In fact, the astronaut wrote that "*it would take three days*" ;

without explaining why the delay.

On the way
Next, the spacecraft "*flew into the shadow of the Moon*" ;

without explaining why.
Then the Sun was eclipsed by the Moon, which was "*lit from the back with a bright halo of refracted light*" ;

without explaining what caused the refraction (which can happen in gas or liquid - whereas the Moon has "*no atmosphere*", he also wrote).
Next, "*On the morning of day three, it was time to enter the Moon's gravitational influence*";

apparently unaware that lunar gravity extends *to Earth* (causing tides, for example).
Accordingly "*a burn*" was needed ... ;

without explaining how much fuel used -
... "*to slow* (the last speed he had given was 40,000 kph) *to* (about) *6,000 kph,* and to *swing to the Moon's far rugged side - never seen from Earth*" (despite alleged lunar rotation) ;

nor how to avoid a *"gravitational slingshot"* (see "Google", and especially the illogical comparison with a tennis ball, and remember (1) the non-rotation of the Moon (as almost similarly for Venus) and (2) that the non-rotating lunar gravity would act only centripetally ;

nor why the alignment had apparently not been achieved earlier when firing *"small guidance rockets to check and correct course"*.

Getting to site

Now the craft got *"captured"* (his inverted commas) in thirteen lunar orbits ;

without explaining why so many cycles ;

nor what force continuously <u>changed the heading</u> from <u>straight</u> (as per Newton's first law of motion) to the <u>curve</u> of lunar orbit (at nearly 6,000 kph) ;

and

not referring to the Moon's non-rotation ;

thus

not aiding orbit.

Then *"the time came* (while on the far side) *to separate* the Lunar Module *from the Command Module"* ;

without explaining how.

Next the Lunar Module orbited again, and started a second time, until *"a burn"* enabled *"a coasting descent"* (seeing *"the surface of the Moon rolling by"*) ;

somehow, and again without saying how much fuel was used.

Progress

Meanwhile, Mission Control still *"monitored all aspects of progress"* with *"constant radio contact"* ;

without explaining how those could be done during the time contact was *"**im**possible behind the far rugged side"* (could the same problem be expected behind Earth too ?).

There was also *"static in headsets"* ;

without explaining what caused it (the Command Module had **none** *"fifty miles overhead"* [it had been orbiting *at* sixty]).

Sensationalism

In place of *explanations*, both gossip and high drama were recorded throughout - apparently for the reader's breathless entertainment, such as to report the spellbinding *"twenty seconds of fuel left"* for landing and not having a sensational crash ;

lacking clarification of how fuel provision had been calculated in the first instance ;

and of how a different landing site had been *"planned"*.

Arrival

After touchdown, *Buzz* [**somehow**] changed wine into blood and (through static?) asked the world to *"give thanks for the events of the past few hours"*.

At the arrival stage *"more than anything - they wanted to get out there"*, and, contradictorily, they took their *"sweet time"* to do so ('they' included a fellow-astronaut, *Neil Armstrong*).

"Seven hours after landing" - they *"were ready"* to go out there, and *"Neil opened the hatch"*.

A few minutes later it *"wouldn't release"* for *Buzz* ;

without explaining how *"a tiny bit of oxygen pressure"* [sensationally] prevented opening, with *"the gauge eased down to zero"* (note that scientists have said it's **impossible** to achieve zero mechanically) ;

nor why the hatch had been closed and the craft recharged with oxygen ;

nor how exit was subsequently achieved.

73

MOONSHINE

Cooling

'Out there' had been devoid of any solar influence for fourteen consecutive (terrestrial) days just recently - by the end of which time the ground temperature would possibly have been towards minus some hundreds of degrees Celsius, all around (re-adjustment to the absence of solar influence - not *"heat escapes into space"*).

Subsequently - apparently not more than a tenth of that time later - the Sun (at a low angle - as reported, **ten degrees**) was likely to have induced only a little upward change of temperature (somewhat comparable to early morning on Earth) yet *"with plenty of heat from the Sun and cold in the shadows"* ;

without explaining how temperature was taken ;

nor where there were shadows other than their own and that caused by the Lunar Module (only since very recently).

However, they had *"ice* **water** *being produced by their backpacks* and *circulated in their underwear"* ;

74

without explaining why their suits were apparently inadequately insulating - (during "*a mere two and a half hours*" outdoors ;

nor how refrigeration could take place - in the apparent absence of a fridge's normal cooling by circulating air, of which the Moon had none.

Lighted terrain

Despite the ten degree angle of the Sun, a photo (the "*visor shot*") of *Buzz* shows the shadows of his legs at about **forty-five degrees**, both in the vertical and horizontal planes.

That was **almost** the same as with the Lunar Module (also depicted in a photo), just a short distance in front of him.

The site was on the "*Sea of Tranquillity*" - a few degrees east), and with only very slow lunar revolution to change the low sunlit angle.

Although the background terrain is therefore the same for both photos (all equally as good as '*artists impressions*' in other books *!*), one of them shows it as white to the horizon, and the other grey becoming darker with distance ;

without explaining how only one wavelength of their reported "*monochromatic hues*" could achieve such variety.

75

That terrain is practically <u>level</u> and <u>smooth</u> ;
without explaining why different from Neil's
view out of his window - reported : *"with a*
fairly large number of craters of the five to fifty-
foot variety, and some ridges which are small,
twenty, thirty feet high, and literally thousands
of little one-and two-foot craters" with *"some*
angular blocks probably two feet in size".

Shadowed Moon dust

In two other photos - allegedly   taken in
sequence -   of a patch (for *"a single footprint*
shot"), one (of a boot just *"slightly away"*
immediately after making a print) is sombre-
grey, and the other (of the resultant print) is
bright-grey : both the same *"monochrome"* [!].

In the 'making' photo, the shadow of the boot
at the toe is at **right angles** for the horizontal
plane and of a length suggesting **less than**
twenty degrees in the vertical plane (that is
almost the same for a small bump about a third
of the boot's length in front).

In the 'resultant' photo the bump shown is
more   than the boot length ahead, while the
shadow in the heel-print (impressed as if from
walking, not merely placing) is not only **much**

76

longer proportionately but is at *an angle only half* that for the toe.

Moreover, the *"reverse moulds of the treads"* are clear-cut - in what is alleged to be dust as *"fine as talcum powder"*, which would normally *not permit vertical sides to be sharp at the top*, as in the photo.

Indeed, Neil had reported *"I only go in* (to the dust) *maybe an eighth of an inch"* ;

whereas the 'resultant' print suggests up to four times that (made with weight from <u>lunar gravity only one sixth that of Earth's</u> !)

Solid material

In due course the Moon's *"hard surface"*, and *"soil that wouldn't compress"*, both impeded the erection of a *"hollow flagpole"* - which *"finally we secured."* ;

without explaining how.

"Outside the thin wall" of the spacecraft - both during the three-day flight and on the Moon - there was *"a* [sensational] *deadly, vast, airless <u>vacuum</u>"* ;

without explaining how, therefore, a *"<u>solar wind</u> "* could be so <u>loaded with</u> *"ions of helium, neon, and argon"* that even their infinitesimal

77

"foil-flag" lunar experiment collected *"particles"* - (later described by physicists as *"dangerous radiation"*, also coming in counter-direction from the stars) - in the '*vacuum*' extending *150,000,000,000 metres* from the Sun *!* (and *throughout space* - despite solar gravity).

Indeed, if those gasses had been arriving for billions of years, wouldn't they have formed an atmosphere for the Moon ?

Warming

Before the lift off from Earth, the *"giant rocket"* had been charged with two thousand tons of *"liquid oxygen, liquid hydrogen, and kerosene"* - causing *"large shards of frost already falling off the outer skin"* ;

without explaining the nature and extent of insulation then ;

nor how fuel was prevented from warming similarly during the trip, with *"plenty of heat from the Sun"* (on the Moon - therefore probably elsewhere).

Yet after returning from the two and a half hour lunar walk *"it was* [sensationally] *awfully cold in the lunar lander"* ;

without explaining why *"the completely exposed plumbing"* apparently wasn't <u>frozen</u> ;

nor why, apparently too, it hadn't been cold during the seven hours inside, before the walk ;

nor how they had managed to be *"comfortable"* in only *"flight suits"* during the three-day voyage in space ;

nor what power enabled them to run *"the <u>water</u> circulation system in our suits to warm us"* (earlier, ice water had been needed !) and to *"turn the heat full up* (ineffectively *!*) *inside the cabin"* for *"several hours"* - but apparently not at other times, nor elsewhere (remember, there was *"plenty of heat from the Sun"* [to which the lunar module was continuously exposed]).

EARTH LANDING

Lift-off from site

<u>*Despite*</u> the flow of *"the Sun's electrically charged particles"* (flowing at *"near the speed of light"* [not in *"cruise mode"* ?]) there was *"no atmosphere resisting"* the getting once more *"into orbit"* with *"the right speed* (given earlier as about 6,000 kph) *sixty miles high"* (the *"first of two orbits"*).

In order to do so, Aldrin changed a (<u>*somehow-available*</u>) *"felt-tipped pen"* into a *"circuit breaker switch"* - to avoid another [sensational] disaster.

All these feats - plus coming *"up from below"* to *"the rendezvous and docking"* with the Command Module, *"nearly four hours after lift-off"* - were *"absolutely beautiful"* ;

thus ***evading question*** by the reader - such as whether there was more than " *twenty seconds of fuel left* ".

Then, after docking, the Lunar Module was discarded.

On the way back

Next, to *"line up to the Earth"*, *"Mike Collins*

(driver of the Command Module) *guided us into the Trans-Earth Injection burn, the extra push* (the force required by *Newton's* first law) *that would consume five tons of propellant in less than two minutes"* ;

*sensational **boozing** !*

That consumption might be comparable with Aldrin's wine-drinking and gushing thanks - this time ***sixty bottles per second*** for more than one hundred seconds ;

without explaining what sort of plumbing and equipment could cope with such a flood.

Whatever the reason, its alleged effect was to "*boost our speed by* (about) 3,000 kph", and "*break us free of the Moon's gravitational pull* "

overlooking the evidence that lunar gravity influences as far as Earth, mutually.

Fuel dependency

However, in the likelihood - that there is a direct ratio between amount of fuel and the (gravitational) force to be overcome by it (*Newton* again) - the "*five tons of propellant*" and the "3,000 kph" would be related to the mass of the Command Module.

Apparently it had twice the mass of the Lunar

Module, so the latter's earlier lift off, into orbit at (about) <u>6.000</u> kph, would similarly require five tons of fuel (plus the same for its earlier descent).   Such amount would therefore require tanks with capacity of approximately ten metres cubed (compared with the measurement of water, and with the liquid gas of the *"giant rocket"* - more than one hundred metres tall and with a ten-metre diameter - for *"more than 2,000 tons of propellant"*) ;

nevertheless, the photographic impressions and descriptions of the Lunar Module (powerful engines, much equipment, and working space for two astronauts during two days) seem far short of also having great tanks equivalent to the extra space of a single bedroom.

Later, the discarded Lunar Module *"crashed on the Moon"* (how known ?) *"after its fuel and batteries ran out"* ;

without explaining why   those two power sources should be necessary to the continuation of orbiting (but apparently not previously during the trip - of which there had been several circuits).

Travel time - back
   Subsequently, the *"three-day* (72 hours)

journey back to Earth's upper atmosphere (which apparently is very rare) *was relatively uneventful"* ;

without explaining the difference between that long time and the 44 hours (approximately) needed at a speed which was about 9,000 kph (lunar orbit plus the boost) when allegedly leaving the Moon.

Next, *"when the Command Module hit the Earth's atmosphere we would be travelling at over 25,000 mph"* (40,000 kph) ;

without referring to the implications of Galileo's falling formula, which relates (vertical) distance to the square of time (periods) taken - just calculate *that* geometric progression (starting at 9,000 kph) ;

nor explaining that the 44 hours would therefore be greatly overstated without some *explanation.*

More sensationalism

By this point *"the Service Module had been discarded along with our remaining fuel to manoeuvre our craft"* ;

apparently no *"guidance rockets"* ;

without indicating how to *discard* ;

nor why ; and

without indicating how to avoid entry (at 40,000 kph) *"too steep"* with *"intense heat"* that would [sensationally] be *"fatal"* ;

nor how to avoid entry *"too shallow"* and being *"deflected"* - therefore [sensationally] to *"shoot off into space"* where they would *"run out of fuel"* ;

nor how alleged deflection (another mere invention ?) by <u>rare</u> atmosphere could overcome - in the craft - the enormous Galilean inertia induced cumulatively by Earth's immense <u>centripetal</u> force (thus impossible to escape without *"a burn"*, as had been necessary on approaching the Moon) ;

nor how, even, they could run out of something they hadn't got.

Conclusion

Didn't *Buzz* take his readers for quite a ride, and haven't some renowned physicists done likewise ?

AIDE-MEMOIRE

Acceleration 1, 32, 41, 65   Aether 67
Agitation 18, 22, 48, 49, 51   *Aldrin* (Buzz)
69 to 84   Ampere 15   Aphelion 42
Archimedes 31   *Armstrong* (Neil) 73 to 77
Aristotle 29, 35   Atoms 14, 17
Attraction – see 'universal gravitation'
Awareness-mechanisms 18, 23

Bent 46   Bicycle 45, 59   Big bang 11 to
13, 27, 47, 50   Black holes 78, 148
Bruno (Giordano) 35

Celsius 13, 52, 74   CERN [Conseil Européen
pour la Recherche Nucléaire] 20 CO_2 21
Cosmological constant 66   Creation 18
Cruise mode 61, 63, 80

Descartes (René) 6, 8, 40, 67
Downward 42, 59

Earth 12 *et al*   Egg-shaped orbit 66
Einstein (Albert) 18 *et al*   Electricity 15, 16

Motion [*first law* - straight line] 40 ; [*second law* - acceleration, inverse] 32 ; [*third law* - equal, opposite] 41 ; [*planetary*] 42 Mutuality 13

NASA [National Aeronautic and Space Administration, of the U.S.A.] 61-2   Neutron 16-17   *Newton* (Isaac) 40-42   *et al* newton 15   Nonsense 23, 35, 57, 58

Perihelion 42, 65   Photon 43-44 *et al* Planets 42, 60,64, 66   *Planck* (Max) 43 Pressure 48, 73   Proton 16-17, 49, 67 *Ptolemy* 35   Pulsing 15

Quantum 43   Quark 17

Rabelais (François) 67   Ray 43, 67 Reaction-time 20   Receipt 57

Simultaneity 55-57   Singularity 28

Relativity 28, 34-35, 46, 48, 55   Rest 30, 40 Revolution 22, 37, 62, 75   Rotation 22, 27, 37, 61, 65, 70, 71

Shakespeare (William) 34 46-47   Space-time 28,

Thermos 68   Time 21   Titanic 53
Touch 48   Trajectories 37

Universal gravitation [*law of*] 23

Vacuum 67, 68, 77, 78   Virtual 43, 44
Volcanoes 49

Wavelength 45, 75   Weight 29, 44, 53, 77
Wells (Henry George) 63   [author of *The First Man on the Moon* (1901) / in a repartee *Jules Verne* bantered *"can Mr Wells show me some cavorite."*]   West Bank 58

88